Revelations Chapter 9–Verses 10 - 11

"... and there fell a great star from heaven burning as it were a lamp, and it fell upon the third part of the rivers, and upon the fountains of waters; and many died of the waters, and the name of the star is called wormwood..."

Chernobyl is the Ukrainian word for wormwood.

Chapter 1: Introduction

Russell Crockett was born on a cold December night in remote Shady Valley of the Appalachian Mountains. Winters in the high country of Tennessee can get miserably cold especially in a meager farm cabin such as the one in which Russell was born and raised. Russell was the youngest of four Crockett children each born in the little cabin. Two of his siblings, a brother and a sister, had died during their first year of life. Russell nearly did so himself barely surviving successive bouts of whooping cough, measles, diphtheria, mumps, and chicken pox. Perhaps his survival may have been due to his inheriting more than his share of the tough constitution of his pioneer ancestor, Davy Crockett. At least, now that he is older, Russell likes to believe that to be the case.

Shady Valley was completely surrounded by a ring of mountains–Watauga to the north curving eastward and the Clinch to the south curving westward. The only tie to "civilization" was a winding narrow mountain road, impossible to penetrate during most of the winter. The nearest town was Elizabethton which was aptly named because the colloquial language in this isolated area more nearly resembled Elizabethan English than anything else. It was this that colored Crockett's accent making it different from what is often considered southern.

Although reared in the Tennessee back-woods country in abject poverty, Russell remembers his youth not as a hardship but as being idyllic. Poverty defined as the lack of money was of little concern in the Tennessee mountains as there was really no need for money. Mountain people made innovative use of natural resources enough to be completely self-sufficient.

The young Crockett, along with a half dozen other miscellaneous mountain children, was schooled in a one room stone school house that combined first through eighth grades and served as a church on Sundays. Heat was supplied by a cast iron wood stove in the middle of the room vented through a hole in the wall. The older students shared the teaching load of the two permanent teachers. Crockett did more of his share of this because he was blessed with relative "smarts" and had easily mastered the required curriculum.

School was only in winter months. The summers required strenuous work just to scratch an existence out of the tough red Tennessee clay soil. Nevertheless, there was ample free time for other activities. The summers then were particularly special to Crockett perhaps as some sort of inherent compensatory reaction to being born on a cold December night. Crockett spent his free time penetrating as deeply into the surrounding mountains as he could–sometimes spending as much as a week away living off the land (chinquapins, roots, wild berries, nuts, fish, crawdads, and edible greens). By the time he was 14, he knew every trail, every peak and "holler", every stream, and even individual trees for miles around. Although often in unfamiliar territory, he was never lost. He possessed that instinctive sense of orientation and direction common to most mountain folk.
In this environment, Russell grew up hard and lean, self-reliant, but a loner. Lacking

somewhat in social graces, his usual self-assurance was not apparent whenever he encountered any of the half-dozen or so young girls near his age that lived in and near the valley. His usual prideful and even graceful mien turned into a tongue-tied awkwardness in the presence of these delightful creatures that he held in such awe. He had neither a talent nor a liking for the female mode of conversation. Nor did he have any taste for their social functions that mostly revolved around the little church. He didn't feel right about any participation in the purely social activities at the church as he didn't participate in any of the religious services. It was not so much that he didn't "believe". He actually wasn't quite sure of what he believed about God and heaven and such. It was just that he did not give these much thought and he did not like being fettered by the excess baggage that went along with religion. And, besides, Sundays constituted prime time for his excursions into the mountains.

As Crockett was finishing the last day of his tenure at the school, the schoolmaster, (Mr. Tate) asked him to stay a few minutes. "Russell, you are a very bright young man. I've never had another student that was able to so quickly and easily master what little math and science I've been able to cover here. Do you have any plans to continue your education?"

"Ah don't rightly know how ah could. As you know, us Crocketts are sorely lacking in money."

Mr. Tate said: "I've known about a very good small college in Kentucky that will take students like you without tuition. They will give each student an on-campus job to earn their keep while they get an education. This school was established exactly for people like you -- promising Appalachians that, after graduation, might return back into their region as teachers to perpetuate the process - an interesting regional boot-strapping concept. If you are interested, I can tutor you on the high school requirements and later administer the entrance exam."

Crockett was elated: "Ah would be most appreciative if you would do this favor for me."

So, it was only about a year later that Crockett showed up on the campus of the small liberal arts school. Younger than the other entering freshmen, still shy and a bit socially backward, he didn't fit into the social life too easily. It was not long then before he acquired his current nickname of "Tennessee". He was labeled so because of his effusiveness when comparing his home area with other places to live, but, more so because his Appalachian twang differed from the others and his clothes were of a more peculiar cut. Although the nickname was good-heartedly derisive, Crockett almost eagerly accepted it as an honor, for, if the truth be known, he never really liked his given name.

It was natural that he chose Physics as his field of study because Tennessee was possessed of a burning curiosity about the world and its working. He saw the study of Physics as being the true study of philosophy. He viewed with disdain the pap that was fostered off on students in the name of philosophy under the aegis of the Philosophy Department and wanted no part of that.

The exposure to the "more civilized" environment at the small college did much to smooth out his rough edges. He even found that occasionally he could carry on a sustained conversation with a pretty girl – providing of course that she had initiated it in the first place. Oh, there were still frequent lapses such as a propensity for tripping on virtually every up stairway. And, there was the time he stumbled over an imaginary crack in the boarding hall floor, scattering the contents of his loaded food tray onto a table occupied, naturally, by several coeds who were undoubtedly the reason for his being distracted in the first place. It was not that these were isolated incidents either. Such seemed to happen to Tennessee with astounding regularity.

So, it was with some relief when he graduated with his BS physics degree and accepted an entrance level position at the prestigious National Laboratory in Oak Ridge -- back in his beloved Tennessee. Once there, however, it didn't take him long to realize how little technical knowledge he had compared to the 900 or so PhDs at the Oak Ridge Laboratory. He was concerned not so much about potential advancement but more so about his ability to do good science and to better fit into the Laboratory environment. As there was little chance of him ever being mistaken for a Harvard graduate, he decided that he must further his science education. Characteristically, he met this challenge head-on the only way open to him -- that was to continue working days but studying in night school at the nearby university. Having seen first-hand the devastation wrought upon his beloved mountains due to clear cutting and strip mining, he became convinced that the forests and the mountains needed his help. Consequently, he concentrated his studies on nuclear engineering. To him, the hard grind of day-work and night-classes were just another mountain to climb. Thus, after more than the usual number of years, he eventually became an unlikely crusading hillbilly PhD nuclear engineer who still had an Appalachian twang and a propensity for tripping over his own feet – a strange creature indeed

Chapter 2: The Announcement

It was Wednesday, April 26, 1986 that Tennessee took some time off from work to spend the day doing his favorite thing – trout fishing on a Smoky Mountain stream. This time he had chosen Abram's Creek in the foothills that surround Cades Cove. At the present point in the stream, almost half way around the six-mile stretch known has Horseshoe Bend, there are seemingly un-scalable cliffs that rise sharply on either side. The entire area seems to be covered in a velvety virgin green carpet that is characteristic of an early Tennessee spring. Here the stream narrows as it squeezes between two huge boulders markedly speeding up in velocity in dutiful obeisance to Bernoulli. A few yards further down, the stream drops sharply into a pool, creating a myriad of white frothing bursting bubbles. Just out of reach of the spray, a new hatch of willow flies re-enacts their age-old genetic imperative process of fliting up and down near the water surface. Occasionally an ephemerid will land on the water to become fish food thereby playing out the peculiar drama of life as a small link in the food chain.

One fly drops slightly too heavily onto the water and seems to linger there. Beneath the water, the "old one" fixes his one remaining good eye on the now floating morsel and readies himself to charge at just the right moment. But he hesitates. There is an almost imperceptible strangeness in the wake. He instinctively lets this one pass. No wonder he has survived to become the "old one". A less discriminating young brook trout shoots upward and with a delicate twist sucks in the prize. Immediately as the trout breaks the surface, Tennessee flicks the tip of the fly rod and sets the hook. By deftly "fingering in" the line, he maneuvers the trout into a shallow pool. He gingerly lifts the fish from the water momentarily admiring its shimmering beauty–an almost perfect torpedo shape with tiny brown spots on the side that blend into brilliant orange patches on the bottom. Crockett tenderly removes the hook and returns the trout back into the stream. The trout, without so much as a thank-you, darts back into the pool depths to sulk at the insult, but nevertheless wiser for it all. This is the eighth time today that Crockett has repeated this scenario – just as he has done many times before in his thirty-some years as a native Tennessean.

Taking a much-deserved break, Russell stretches out on a flat boulder in mid-stream to soak in the little sun left.

In nature, there are special sounds that resonate with our deepest chords. One such sound is made by a mountain stream as it submits to the gravitational imperative by plunging along the paths of steepest descent until it receives a rest at the bottom. This indelible sound is what lulled Tennessee into a deep sleep.

A sudden thunder clap arouses Tennessee from his sleep. The thunder and not too far off lightning portends the development of a sudden late evening spring storm. As the sun has already begun to settle behind the close-in mountain tops, the arriving dark thunder clouds cast an eerie early darkness. For the unawares, a Tennessee mountain storm in all its spectacular electric grandeur poses ominous real hazards. The down-pouring deluge in the upper regions of the tributary watershed can feed the major stream almost instantaneously creating a virtual wall of water crashing head-long downstream with devastating power. Many an unsuspecting fisherman has met an untimely end from these mountain tsunamis. Tennessee himself once had a close encounter with one of these spectacular events and he had no desire for another.

As the darkness was descending rapidly and the developing storm appeared to be on the verge of mounting a major onslaught, Tennessee considered his options. He was too deep into Horseshoe Bend to retrace the trail back. He deemed the quickest route out of danger for him was to climb directly up the cliff side. He had first to maneuver on hands and knees through the thick jungle-like mountain laurel that grew contiguously by the stream until he reached the base of the cliff. As he had anticipated, the climb up the cliff was not difficult. He knew of an ancient trail used by the aboriginal Cherokee Indians that had been kept passable by several animal types. Taking this trail, he was relieved that it was not yet fully dark when he reached the most treacherous stretch … a narrow rocky path undoubtedly cut into the cliff side by the Cherokees as an easily defendable route of escape from enemy tribes. Slightly wider than a foot, this stretch of the trail had cliffs rising vertically on one side that dropped off perilously for hundreds of feet on the other side. If the oncoming night were any darker, this section of the trail for the next mile would be difficult to traverse. At times like these, Tennessee's natural awkwardness transforms into the sure-footedness of a mountain goat. He was making good time along this stretch until he encountered one of Mother Nature's perverse practical jokes. A large boulder had slid from the cliff side and had unaccountably lodged itself directly in the middle of the trail. Tennessee used the knife he always had with him on such outings to gouge hand and foot holds into the side of the cliff. With these, he was able to climb onto the top of the boulder. Hanging his legs over the other side, he prepared to jump back onto the trail. Before he could do so, the high pitched ...zzzzzzz... froze him into place. This was the unmistakable warning of a Smoky Mountain diamond--back rattlesnake. Lying on the ledge just below the boulder, the reptile was awaiting a much-needed dinner. His sensitive infra-red sensors, however, had indicated that "Tennessee" was "choke-size" and would not make an appetizing meal anyway. Thus, the ...zzzzzzz... was his way of saying "stay out of my territory, big boy, or you will be sorry."

Tennessee's first impulse was to leap over the snake onto the narrow
trail and rapidly continue forward. This looked chancy given the narrowness of the trail. Another option was to just stay put and weather the storm until, hopefully, the snake would go away. Staying put, however, was never something that appealed to Tennessee. Still on top of the boulder, Tennessee noted that he may be able to reach the rattler if he could anchor a foot and hang upside down. He locked one foot into a handy root in the cliff side and leaned

down toward the quarry. He gingerly brought his right hand to within about two feet in front of the curled snake and slowly moved it back and forth in rhythmic fashion. The rattler fixed his vertically pupilated eyes rigidly on the undulating hand – poised to strike at any moment but seemingly hypnotized by the movement. Tennessee deftly slid his left hand in from the back side and with cat-like quickness grasped the snake firmly on the nape of the neck. With one continuous motion before the snake could react and coil itself around his arm, Tennessee flung the rattler over his right shoulder. The rattler landed with a dull thud on the downstream of the trail some 5 yards in back of Tennessee. It quickly re-coiled in anticipation of a second assault which never came. Tennessee was already proceeding over the rest of the trail – a wide grin decorating his rugged face.

The dalliance with the snake was too long. By the time he reached another trail that would lead him back down the mountain side to his parked truck, darkness had completely settled in. In the Smoky Mountains, light has a difficult time to penetrate the high peaks soaring above the trails which wend their way beneath great tall trees. On such a night with no moon and all starlight blocked by clouds of an impending storm, the darkness can be absolute. The trite phrase "you can't see your hand in front of your face" becomes literally true. It is a condition of zero visibility.

Tennessee had only been caught in the mountains like this once before as a result of his having too much fun fishing and thereby failing to start back early enough. That time was during mid-summer when "lightning bugs", as fireflies are called in East Tennessee, were in force. One of Tennessee's most vivid memories is of that summer night years ago when these little miniature lanterns that were stirred up with each step produced sufficient light for him to easily follow the trail down the mountain–an almost mystical experience which left Tennessee with a different view of these insects – in a different light so to speak. This time, however, there was no such luck. Tennessee slowly and meticulously made his way step-by-step using only his sense of sound and touch. Each lightning stroke of the now arrived storm momentarily showed him the lay of the trail for a few yards ahead. The storm's rain was fierce and cold and was converting the trail into a temporary stream. Slipping and sliding his way with each lightning stroke, Tennessee's progress was extremely slow. Fortunately, just as quickly as the storm arose, it suddenly abated and temporary quietness settled in. Usually on a spring night, there is a virtual cacophony of sound emanating from creaking trees, falling branches, katydids and a myriad other insects. However, this night, perhaps a byproduct of the storm, things were eerily quiet and now there was no more lightning to show the way. Tennessee could hear his own footsteps echoing in the forest – in his imagination sending out clear signals to whatever entity that might be lurking in this total darkness. Feeling vulnerable, he was momentarily taken aback when he noted a ghostly apparition a few yards off the trail to his right. Whatever it was momentarily sent shivers over his whole body as it lay there shimmering in a cold white light. After the initial fright, Tennessee immediately recognized it. "Wormwood" Tennessee blurted out. "It's wormwood." Tennessee remembered that rotting southern hemlock is attacked by a certain type of rare fungus that is naturally florescent and emits a faint white light, imperceptible in daytime but clearly visible at night. Among the superstitious mountain people, the infected wood is considered to be an evil presence to be avoided at all costs. A chance encounter at night is believed to mark the unfortunate for a visit from evil spirits. Tennessee had no such illusions. However, as he picked up a large piece of the glowing wood he could not fully suppress a shudder. Perhaps being rescued from a difficult situation by this rare find may actually be some sort of omen. Holding the dim light in front of him to light the way, Tennessee quickly covered the remaining three miles down the trail to the welcome sight of his 15-year-old fire-engine red Dodge pickup truck.

To return back to Oak Ridge, Tennessee decided to avoid Gatlinburg and Pidgeon Forge and take the back route through Townsend. The rain had restarted and was beating steadily on the windshield. This didn't bother Tennessee as he liked to drive at night in such a rain. He was thoroughly relaxed as he listened to the Sevierville country radio station playing an appropriate Ronnie Milsap song:

> *"Smoky Mountain rain keeps on fallin'*
> *I keep on callin' her name*
> *Smoky Mountain rain I'll keep on searchin'*
> *I can't go on hurtin' this way*
> *She's somewhere in the Smoky Mountain rain"*

The radio got his attention with an interruption. "We interrupt this program to bring you a special news bulletin. Swedish scientists have reported measurement of an intense cloud of radioactivity passing over their country. It has been reported that the activity is emanating from a Russian Nuclear plant that has experienced an explosion apparently last night and is now on fire and is spewing radioactive material high into the atmosphere."

"My god" Tennessee gasped. "I'd better get back to the Laboratory.

Chapter 3: Preparation

Tennessee was in his office early the next morning pouring over what little information he had in his files describing the RBMK reactor type thought to be the one having the reported accident. Scanning rapidly, he read, "... graphite moderated ... positive void coefficient ...," "Holy Shit, those turkeys have a positive void coefficient like our discontinued fast breeder reactor concept!"

Tennessee had spent a large part of the preceding few years evaluating the safety aspects of the proposed sodium cooled breeder reactor that was to have been built near the Oak Ridge reservation almost in his own back yard. While the sodium cooled concept had many safety advantages, it had one outstanding flaw. If for some reason the liquid metal coolant were to be rapidly removed from the core, the effect would be to markedly increase the power level rather than to decrease it – the opposite of the effect that would occur in the water-cooled reactors now in operation in the U.S. and elsewhere. The reason for this behavior is that the sodium absorbs neutrons and makes some of them unavailable to fission the Uranium fuel. With the sodium removed, there are so many excess neutrons available that the reactor goes super-critical on just the prompt neutrons alone (some fission-released neutrons are delayed slightly). The significance of being super-critical on the prompt neutrons alone is that the rate of power increase is very rapid and uncontrollable by inserting the scram rods. The excursion can then only be turned around by either of two ways: (1) the temperature of the fuel increases to such a level that it no longer effectively absorbs neutrons (the Doppler effect), or (2) the fuel itself flies apart due to the rapid generation of internal pressure.

When Tennessee was involved in assessing the safety of the fast breeder reactor, the "hypothetical" accident of most concern was called the LOF-driven-TOP. In the jargon of nuclear safety, this means a Loss-of-Flow (sodium voiding) driven Transient-Over-Power accident. This is a euphemistic way of saying that the reactor blows itself apart. The major safety debate with respect to the LMFBR was whether or not the LOF-TOP produced enough energy to break open the heavy pressure vessel that housed the core – and, if it did, would it also create a leakage path through the containment to the environment.

Tennessee was convinced that a resultant sodium fire would create such a leak causing release of radioactivity to the outside environment. This is the reason he was taken aback by the revelation that the RBMK type reactor also had a positive void coefficient.

Tennessee's concentration was interrupted by the appearance of someone in his doorway. "You're in bright and early Russ. I'll bet you are thinking about that Soviet nuclear accident. What's it all about anyway?"

Katie, the young secretary assigned to the Nuclear Safety Evaluation Division, always addressed Tennessee as "Russ". She was also a native East-Tennessean so she overlooked his peculiarities and the nickname didn't really make much sense to her. Katie was not technically trained, but she possessed a sharp and eager mind. So, whenever she expressed interest about technical matters, Tennessee always tried to oblige and explain things in layman's terms as best he could.

"I don't actually know yet what set it off, but I'd bet my coon-skin cap it involved the positive void coefficient."

"Well par-ar-dun me, Mr. Scientist. What is a positive void whatchamacallit?"

"You recall that a nuclear reactor gets its power because each atom of U-235 in the core absorbs a neutron and undergoes a fission reaction that releases both energy and also $2^1/_2$ additional new neutrons on the average." Remembering his early days of tutoring the younger pupils in the little school house in Shady Valley, Tennessee was now getting into this–feeling much like a university professor.

Katie, somewhat amused at his intensity, interrupted. "What do you call half a neutron: a neut or a tron?"

Tennessee smiled at her humor. "There are no half neutrons–just $2^1/_2$ on the average -- sometimes 2, sometimes 3. If each of these were to be absorbed by another U-235 atom which would then fission and produce $2^1/_2$ more–and so on–there would be an uncontrolled chain reaction."

"Like an atom bomb?" Katie asked a little more seriously this time.

"Somewhat", replied Tennessee. "But in nuclear reactors this is prevented from happening because some of the neutrons leak out and some get absorbed in materials that do not fission. That's why control rods are put into a reactor. Their non-fissionable materials have a strong affinity for absorbing neutrons. These absorb the excess neutrons so that, on balance, each fission leads only to one additional fission thus keeping the system in steady state – a condition we call just-critical."

Before Tennessee could start again, the telephone rang. As he lifted the receiver and leaned back in his swivel chair, the coiled coffee cord swung wide across his desk and sent his just emptied coffee cup crashing to the floor creating a rosette of shattered ceramic fragments. Tennessee was used to such occurrences by now and, without so much as a double-take, he answered "Tennessee Crockett here."

"Hello Tennessee, its Vic. Can you come over to my office right away? We want to get together on this Chernobyl thing."

"Sure thing," Tennessee replied. "I'll be there in a jiffy."

Victor Parnelli was a vice-president of the Laboratory and was Tennessee's immediate superior in the organizational hierarchy. Tennessee liked Vic because he had made it to that level in the organization with his technical competence and not through some artificial managerial style. Vic had made a name for himself during the TMI-2 accident, being one of the few "experts" early on the scene that, before the facts were fully known, had provided a reasonably accurate interpretation of the events as they were unfolding. This was particularly true with respect to the "hydrogen bubble" problem produced as a result of steam reacting with the high temperature Zircalloy clad around the fuel. Tennessee knew that Vic had paid his dues over the years and had made many valuable contributions to the nuclear energy field.

Tennessee's punctuality followed its usual course – he arrived after the meeting was already underway. Vic was already orchestrating the proceedings. He was directing Varnes Parsons, a pioneer in the field who had been around the Lab since the Manhattan Project days, to get in touch with his foreign contacts in Sweden. "See if you can get detailed information on the quantity and the isotopic mix of the fallout they are measuring. Look particularly at the ratio of Cs-134 to Cs-137 and maybe you can pinpoint the exact time the accident started, and maybe give us some idea of whether it was a production or a power reactor." Vic then turned to Huffstet, an atmospheric transport specialist, and directed him to start gathering the data on the meteorological conditions and to use his plume models to see if he can make a "back" calculation from the Sweden data to estimate the source strength.

Lloyd Olsen of the Analysis Section was asked if he could adapt and exercise his severe accident computer code to model the accident itself when the details of the scenario are known. "You could start by modifying the code so that it is applicable to the RBMK type of reactor."

Olsen replied, "It won't be easy – it's almost like re-writing the whole code because the geometry and nuclear characteristics are entirely different from our reactors."

"No matter," said Vic. It will have to be done sooner or later and we might as well get started on it now."

Tennessee could see that Victor was getting the Lab into full mobilization for this particular crisis, just as he had done for TMI-2. Tennessee was genuinely impressed during that time with the ability of a National Laboratory to rapidly mobilize a wide range of expertise and capability and bring it to focus on a national problem. He couldn't help but reflect on what real national treasures these labs are and how lucky he was to be working at one and still be living in the midst of his cherished Tennessee mountains. One fleeting thought, however, was hard to suppress. The nuclear experts themselves seemed almost perversely pleased that something exciting had happened to give them the opportunity to exercise their wits on something other than the sometimes-tedious research activities. They were as energized as Tennessee had seen them to be at any other time since TMI-2.

Vic turned to Tennessee. "Carl Zander of DOE called. He wants me and you up there on Monday morning so he can pick our brains on this accident and see what they need to be doing at this stage. He tried calling you first and when he couldn't reach you he called me. Dammit, Tennessee, it took me twenty minutes to convince him that I also know a little something about reactor safety. He insisted that you come to the meeting and, by-the-way, I could come along too if I wanted."

"Ouch," Tennessee winced. "Sorry about that, Vic. When we see him, I'll try to explain that you're not just another pretty Harvard face."

The plane from Knoxville to Washington goes via a stop and plane change in Atlanta. Vic noted "when I die and go to heaven, I'll probably have to go through Atlanta to get there." Tennessee said, "Maybe, but when I die, I will probably transfer to a different plane than the one you take."

The plane from Atlanta lumbered into Washington National early Monday morning. As they made their way toward the rental car desk, Vic wore a disgruntled look. "Look at all the tax-payers' money they spent to spruce this place up. The only reason it exists is for the convenience of Congress so they won't have to drive out to Dulles for their week-end junkets back to their home state so they can play politician." Tennessee agreed. Personally, he liked National the way it was before it was refurbished.

Later at the rental car desk, there were about half a dozen impatient business men in line ahead of them. Several were dressed in the current traveling uniform – the ubiquitous blue blazer, tan pants, and brown wing-toed shoes. Whenever visiting Washington, Tennessee does pay homage to the unwritten dress code with one exception: he still wears his boots as his way of showing he is still a mountain man.

A dark haired young clerk with a badge over her left breast that read "Margarita" shot staccato questions at Vic ... "work phone?" ... "residence" ... "where staying while in town?" ... etc., etc., etc.

Tennessee thought: "how unnecessary in this day of computers. It is not good to keep businessmen waiting too long. They get very impatient after about 5 minutes. One of these days, a sharp company will do away with this process and have the rental car waiting as you arrive with no additional information needed."

Tennessee had made this trip so many times in recent years that the process of picking up the rental car, the drive around the belt-way and then to headquarters was almost automatic. Whenever Vic accompanied him, Tennessee usually rented a car in deference to Vic's rank rather than take the Metro which he preferred. He also did the driving so that Vic could spend the time in route preparing for whatever was on the upcoming meeting agenda.

It was 9:30 a.m. before they reached headquarters and the meeting had already begun. Tennessee and Vic tried to unobtrusively find an empty seat. Tennessee's foot made its inevitable confrontation with the cord of the viewgraph projector. His reaction speed saved him from a personal spill, but the carefully arranged viewgraphs of Carl Zander fluttered haphazardly to the floor amid some poorly squelched laughter from around the room.

Carl Zander finally recovered his aberrant viewgraphs and said, "now that we have had our comic relief, let's get on with the business of this meeting."

When Zander spoke, it was in a gruff voice which seemed proper accompaniment for his massive size. Tennessee guessed there would be about 250 pounds on his six feet four-inch frame. He wore his poundage well. The excess bulk made him seem all the more imposing. Tennessee mused, "if I am ever in a fight, I want him on my side."

Zander had taken over and, like many a governmental agent, he was dominating the meeting. He was in his element here – presenting quickly prepared viewgraphs diagramming the design features of the RBMK 1000 type reactor. Zander slapped the top of his pointer against the screen for emphasis while showing the cut-away image of the reactor system.

"Water is boiled at 14.5% quality in 1661 zirconium alloy pressure tubes – 88 millimeters in diameter. Each of these has its own fuel assembly of 18 fuel elements. These pass up through central holes in 250 mm x 250 mm x 600 mm high stacked graphite moderator blocks. The full core assembly, including the graphite moderator, is 11.5 meters in diameter by 7 meters high."

As Zander reeled off feature after feature of the Russian reactor design, Tennessee's mind started to wander because he already knew these details. He gazed around the room and immediately recognized four familiar faces, James Arnold and Robert Sterling of Columbus Laboratory, Simon Trent III from Long Island Laboratory, and Howard Clemons from Albuquerque Laboratory.

Tennessee suspected these four would be there because, along with Tennessee, they constituted what was sometimes referred to by others as "the fabulous five or the F5 group for short." This group, all PhD engineers, had been pegged by DOE management as unofficial consultants on any issue related to reactor safety and they were often together on any committee, crisis, briefing or whatever whim DOE thought serious enough to bring in their "first team." These five represented a composite personality and a combined expertise that would be difficult to match anywhere else. Tennessee felt very proud to be a member of this group. Although they often had heated discussions, he had come to like each of the four as individuals and to respect them as professionals.

Of the group, the one nearest Tennessee's age was James Arnold sitting diagonally across the table. Graying a little, thinning a bit on top, and sporting his ever-present plain pipe beneath a full mustache, James had a pleasing, sensitive appearance yet distinctly masculine. Tennessee had often observed that James had a certain appeal to the opposite sex and wondered if he himself might have any such similar appeal. He quickly dismissed this idea as not being very likely.

Tennessee liked James' laid-back approach to things. No matter how hectic and unreasonable were the bureaucratic demands, James never seemed to get flustered and even seemed to be slightly amused. Tennessee had on several occasions observed James telling his Washington "masters" to quit badgering him and get off his back-- he would deliver on the day he promised and not on their demanded two weeks sooner. But, then he would quietly return to his home lab and provide the desired product on the earlier date. Tennessee was onto James. His seeming "don't-give-a-damn" attitude was just a facade.

Sitting at James Arnold's right was Robert Sterling also from Columbus Labs. Tennessee had to restrain himself from labeling him with the nickname of "Mr. Princeton." He seemed the incarnation of the description Tennessee once read of a Princeton Man: smooth, immaculately attired, well educated in the arts and humanities, and very well spoken. This image was somewhat unfair to Robert because his refined and reserved persona belied what was actually a warm individual with a wry sense of humor. And, besides, he was one of the best technical scientists of the group.

The third member of the traveling circus group was Simon Trent III of Long Island Labs. Simon was one of the fifth column contingent of the 60's brain drain from Great Britain when a number of top British Scientists immigrated to the U.S. to take advantage of the better opportunities and the much higher pay. Simon, however, had long since replaced his "scientist cap" with a "managerial cap." Nevertheless, he was still a valuable member of the group because of his organizational skills. He was often the leader of the group when cooperative activities were required. He also often served as the liaison with the DOE sponsors because he was more able to speak their "managerese".

Howard Clemons of Albuquerque Labs was the youngest and the most enigmatic of the group: bespectacled, prematurely bald, a heavy smoker, and seemingly having a Napoleonic complex because of his short physical stature. He was, however, clearly the most intelligent and best scientist of the group. When he opined about a technical matter, you had better listen because he would likely be right 99% of the time. He had the unique ability to concentrate on some esoteric mathematical machinations during a technical meeting and yet be completely cognizant of the ongoing activities. Clemons was top rated in his technical education on condensed matter at the top rated technical school in the country: no, not MIT--he went to Cal Tech.

Although Howard was apparently as different from Tennessee as night is from day, Tennessee, nevertheless felt an inexplicable empathy with him and enjoyed interactions with him immensely. Perhaps it was because both were loners and both, for different reasons, were somewhat outside the social norms. The similarity ended there. Whereas Tennessee reveled in the more earthy aspects of life, Howard was on an ethereal plane. While Tennessee was mentally hiking in the mountains, Howard was treading among theories and second-order scientific differential equations.

Zander had finished his description of the known design details of the RBMK-1000 reactor and called for a brief break before proceeding. As Tennessee started over to greet his four compatriots, Zander intercepted him. "I want you to do me a favor. What I want out of your "team" is three things. First, I want to get them started thinking about this accident. Second, I want to hear their speculations as to the possible cause, and finally, I want them to generate as many questions, technical or otherwise, that they can come up with. I want to use these as a basis for getting a U.S. team prepared to go to a meeting with the Russians that is to be held in Vienna to evaluate this accident."

"Sounds like a good idea." Tennessee replied. "What is it exactly you want me to do?"

"I want you to start things off when we reconvene in 10 minutes. Give us your own thoughts about the accident and try to get the group talking. I want a brainstorming session."

Tennessee quickly organized his thoughts and developed a few intriguing speculations to throw out to the group. Tennessee knew he didn't need to say much. In virtually any meeting of any group of physicists or engineers, the problem is not to get them started talking. It is the inverse – how to get in a word edgewise yourself. This meeting proved to be no exception. At the end of the day, Zander had in hand at least six alternative possible explanations for the accident and a list of almost forty questions, the answers to which should pin down the precise nature of the accident and all its details.

The trip back to Oak Ridge gave Tennessee time to reflect on the speculations that were developed at the meeting and to sharpen his own theories as to the causes of the accident.

Chapter 4: Vienna

Back in his office in Oak Ridge, Tennessee continued to puzzle over the accident. "I just don't see how they could lead that reactor into such a power excursion. It would either have to be on purpose or would require several very unlikely human errors." As he pondered this thought, the telephone rang. Katie answered it. "Yes, he is here. Just a minute, please."

"Russ, it's somebody named Dr. Olkenoff calling from Vienna, Austria."

"I'll take it in here. Hello, yes this is Russell Crockett. What can I do for you?"

"Dr. Crockett, I am in charge of the Reactor Operations Division of the International Atomic Energy Agency. As you may have heard by now, the Russian have agreed to meet here in September to discuss their findings as to the causes and consequences of the Chernobyl accident. The IAEA is taking on the responsibility of having this meeting to invite the world's press and scientific representatives to inform the world and to identify any further implications and what should be done about them. We expect to produce one of our famous Technical Documents outlining the findings. In this respect, we want to put together a team of experts in four areas: the accident scenario, the source term, the atmospheric transport, and the health effects. We would expect this team to meet separately with the Soviets for an additional week after the full meeting for the purpose of clarifying the Russian assessment report and to put together the Tech Doc. We are looking for experts in each of the four areas for a total of 12 on the team. There would be two from the U.S. We were wondering if you would agree to serve on this team in the area of the accident scenario?

It took Tennessee about a microsecond to say "yes."

"Wonderful. We will send you details and advance material to study as soon as we can. Oh by-the-way, Dr. Howard Clemons of Sandia has also agreed to be one of the U.S. experts in the accident scenario area."

Tennessee was very pleased to hear this. He couldn't ask for a better analyst than Howard to work with in this area. Between the two of them, they should be able to pin -- point precisely the events that led to the accident.

The Russians had hurriedly put together a temporary report on their assessment of the accident and sent it to IAEA which forwarded copies of it on to Tennessee and Howard. There was, however, a minor problem with the report -- It was written in Russian. Fortunately for Tennessee, the Oak Ridge Lab had an employee whose parents came from Russia. He agreed to attempt to translate the report into English. Unfortunately, the translator had forgotten much of his Russian language and was not technically trained. He had a difficult time with some of the scientific jargon. Consequently, there were many parts of the translation that were jumbled and difficult to understand. For just one example, the translation in one sentence of a "hydraulic ram" came out as a "water goat."

Tennessee and Howard spent the intervening time before the Vienna meeting pouring over the translated report and any other information they could get their hands on related to the RBMK-1000 reactor type. Unfortunately, there was not enough time for the rest of the Oak Ridge and Sandia people working on the accident to complete anything concrete for their use with the exception of Parson who had determined from the isotopic data that the Chernobyl reactor was purely for electricity production and was not used for making bomb material.

The day for the initial travel to Vienna arrived much sooner than Tennessee had wanted. He was still very puzzled by this accident and was not sure about the scenario. From the report, he had learned that the accident was not initiated from normal operations. They were apparently conducting some sort of experiment that went south on them. This might explain why Tennessee was at a loss to see how such an accident could occur to this kind of reactor. He looked forward to the one-on-one discussions with the Russian scientists to get clarification as to the nature of the experiment and exactly what went wrong.

After the first stop in Frankfurt, the plane arrived at the Vienna airport around 10:00 a.m. Tennessee had been to Vienna several times before so he knew the way from the airport via taxi to the Nordbahn hotel where rooms had been reserved for him and Howard by the IAEA.

As usual, the Austrian September sky was overcast and there was a chill in the air. The ride to the hotel skirted the more interesting old town part of Vienna but passed by imposing looking ancient castles and governmental buildings. Tennessee was not very interested as he had been this route several times before at various IAEA interactions.

The Nordbahn hotel was apparently chosen because it was the closest to the IAEA headquarters – just three subway stops away. While the Vienna time was only 10:15 a.m., Tennessee was able to check into his room early. This allowed him to take a much-needed nap to help relieve the jet lag. He was awakened by the telephone. A quick glance at the clock indicated he had slept until 3:00 p.m.

"Hello," he grunted.

"Tennessee, this is Howard. I didn't have any lunch and I'm hungry. Do you want to go see if we can find a restaurant within walking distance?"

"Sure. I'll meet you in the lobby."

The search for a restaurant was somewhat in vain. This was not the part of town where good restaurants were situated. They settled for what really looked like a local watering hole. As they entered, Tennessee sensed that it might be a bad choice for a place to get dinner. The place was empty. No customers. No waiters. No cashier. They seated themselves into a booth and waited. Shortly, someone emerged from the back room of the place. Immediately Tennessee got the picture. She appeared to be an over the hill hooker – 40 pounds overweight, too much makeup, and embarrassingly revealing clothing. Howard spoke to her in German as he pointed to Tennessee.

Suddenly, she became angry and grabbed Tennessee by the shirt, yanked him erect, and pushed him towards the exit. Tennessee, of course, didn't strongly resist. At the door, she shoved him out yelling what must have been German obscenities.

Outside, Tennessee asked a laughing Howard "What was that all about?"

"I was just getting even with you for that time in Oak Ridge when you introduced me to a group of sexy looking females that you identified as singles looking for temporary partners. It turned out that they were really "Parents Without Partners.""

"What did you say to that person to get her so upset?" Tennessee asked.

"I simply pointed out that you had remarked that, if this was on your farm and you had a choice between her and a pig, you would choose the pig."

Needless to say, both Tennessee and Howard went to bed hungry that night.

The meeting at IAEA headquarters started at 9:00 a.m. in the huge main auditorium. Tennessee estimated that the attendance was somewhere between 400 to 500 people from various press, governmental, and scientific organizations around the world. The protocol was for the Russian contingent to make their presentations first (one of the four main agenda items each day) to be followed by questions from: (1) the press, (2) the general audience, and (3) the IAEA team of experts in that order. Although the basic language was to be English, there was simultaneous interpretation in five different languages.

This morning's plenary session was started by an introduction from the Director General of IAEA. Although the introductory remarks covered several areas, the following quote seemed to capture the essence of the meeting purposes:

"As stated in the joint communique issued at the time of my visit to Moscow..., the Soviet side expressed its willingness to provide... information on the accident to be discussed at a meeting of nuclear safety experts, in order to assist IAEA Member States to learn from this accident and thus to further improve nuclear power safety."

Each day of the week-long general meeting started at nine but continued until well after six in the evening. From the information presented in the general meeting and the answers to the hundreds of questions asked, Tennessee put together his own version of the accident scenario. Tennessee wrote the following into his personal log of the meeting:

"The accident happened as a result of an ill-conceived experiment planned to be conducted on the operating reactor. In the event of a loss of offsite electrical power, the plants huge main turbo-generators will automatically shut off and the nuclear plant then relies on starting up standby diesel generators to generate the electricity required to operate the emergency cooling systems for the reactor. Starting up these diesels and phasing in their generated electrical power must occur in less than a minute to properly power the essential elements associated with the emergency core cooling system (ECCS). This fast-required response time is difficult to achieve and is a costly part of the plant that must be periodically tested to assure that it works properly. The designers of this electrical system for the RBMK plants had an innovative concept that, if it worked, would allow some relaxation on these requirements. Their idea was to try to utilize the considerable kinetic energy of the main turbo-generators as they are coasting down to still generate electricity that could be used to power the ECCS for a sufficient time until the diesels can take over after a more leisurely period. For this purpose, they had designed voltage regulating equipment and a scheme to automatically adjust to the changing speed of the turbo-generator and still provide the electricity at the right phase and voltage. The experiment was designed to test this new concept to see if it worked as intended.

The test was to be done with the reactor still operating at its minimum permissible continuous power level (700 MWth). This would give the opportunity to repeat the test with various options in case the first option proved unsatisfactory. Rather than use the actual emergency feed-water pump for which the voltage regulator and power were intended, four of the eight very large main coolant pumps were electrically connected to the new device to simulate the electrical load. All eight main cooling pumps were to remain running at the start of the test which would commence by turning off the power to the four connected to the new device allowing them to begin coasting down. At the same time, the turbo-generators off site power supply would be discontinued to start its coast down. The other four main coolant pumps would continue to operate to assure sufficient cooling of the core at the 700 MWth power level. The electrical output from the test device was to be measured with the plants data acquisition system in order to judge the effectiveness of the device.

Had the experiment been conducted under these planned conditions, it is highly unlikely that there would have been an accident. Mother Nature, however, had another of her perverse practical jokes to play on us mortals. After the test engineers had slowly and carefully lowered the power level to about 1600 MWth, an off-site command was given to hold the power there for a while because there was some (unspecified) urgent need for the electrical power. This hold lasted for about nine hours. Because of this long delay in their plans, the test personnel were impatient to get on with the testing. So, after they received the go ahead, they began lowering the power level to the desired 700 MWth on a much faster ramp than was called for by the test procedures. At the 700 MWth level a global automatic control system was supposed to be set to stop the down ramp. At this point, an operator error came into play– the operators failed to set the control switch and the reactor power continued on its fast down ramp declining to a very low level of 30 MWth.

Because of the fast down ramp, the reactor was caught in what could be called the Xenon trap. In any reactor, Iodine is one of the fission products. With a 7-day half-life, Iodine decays into Xenon. Xenon is a powerful neutron poison that plays a strong role in maintaining the reactor at "just critical". It fortunately gets 'burned out" as it absorbs neutrons, so it is held at an equilibrium level at any power by a balance between Iodine decay and neutron absorption burn out. Because the reactor power down ramp was much too rapid, the Xenon essentially stayed at its 1600 MWth equilibrium level even as the reactor reached 30 MWth.

As there were no longer being generated sufficient fission neutrons, the Xenon stayed at this high level. This was very much like having all of the control rods inserted into the core. Even by pulling all control rods out of the core, the operators could only get the reactor power back up to 200 MWth. Remember that at this time all eight of the main coolant pumps were still fully operating. Under the conditions of low power and excess cooling flow, the reactor coolant was in a thermal-hydraulic condition of near saturation throughout most of the core. Hence, the system was on the "steep" part of the void fraction versus steam quality relationship. Thus, any attempt to either reduce cooling (shutting off coolant pumps) or increase power could lead to rapid void increases in large parts of the core. Large increases in void translates into large and rapid increases in power due to the positive void coefficient.

Not being aware of the treacherous condition they had put the reactor into, the test personnel made the unbelievable decision to continue on with the test. On starting the test by turning off the turbo-generator and four of the main cooling pumps to begin their coast down, the increase in void became the dominant influence on the reactivity and the core was led into a super prompt-critical excursion. This resulted in the core fuel rapidly melting and being forcibly injected into the water coolant. The rapid heat transfer from hot fuel fragments to the water resulted in rapid generation of steam resulting in a steam explosion much like some water heaters of the past have exploded. The explosion breached all barriers spewing the fuel fragments and fission products high into the air above the plant. In addition, the graphite blocks that had served as moderator for the reactor were ignited possibly by a subsequent hydrogen explosion and continued to burn

releasing residual fuel and fission products on a continuous basis."

The purpose of the second weeks meeting between only the 12 IAEA experts and selected members of the Russian contingent was to further develop the scenario and to produce the definitive IAEA report on the accident for dissemination to the rest of the world.

On the first day of the second week, Tennessee and Howard awaited the Russian contingent in one of the smaller meeting rooms. They had been informed that the Russian contingent for this area would be Dr. G. F. Demmen, DR. O Pavlov and Dr. Felicitas Roschiv, all of the Kurchatov Atomic Energy Institute. Accompanying these were an interpreter and an unspecified "observer".

The Russians promptly arrived at exactly 9:00 a.m. The introductions were a bit awkward because neither Tennessee nor Howard could keep their eyes off of Felicitas. They had expected a stereotypical Russian female scientist: stocky, no makeup, and hair rolled up into a neat bun. Instead, they were treated with a striking beauty with wavy ebony dark hair and matching brown eyes. When they conducted the obligatory handshakes, Tennessee held onto the hand of Felicitas longer than the normal handshake time. She didn't seem to mind this at all.

The meeting did, nevertheless, get underway. The IAEA experts had already prepared a score of questions that needed clarification.

Whenever any one of these was posed, the Russian interpreter translated it into Russian. The Russians would then discuss it among themselves and then pass their answer on to the interpreter for translation into English for the IAEA team. The two IAEA experts had no idea what went on in the initial Russian discussions. They noted, however, that the unspecified observer took an active role in the discussion generally using words like "nyet" along with disproving head shakes. This process began to grate on Howard and Tennessee because, from the earlier week's interactions, they knew that the Russians spoke excellent English. They had also begun to suspect that the so-called observer was probably an agent of the KGB. Before posing the next question, Howard said to Tennessee:

"ooday oouyay onay owhay ootay eakspay igpay atinlay?"

With a wide grin in anticipation of what Howard had in mind, Tennessee blurted out: "esyay".

They then spoke earnestly back and forth with each other this way for a while. The interpreter had a very confused look and the observer seemed to be in anguish. Howard then turned to the Russians and posed the next question in English.

The astute Russians immediately recognized what the IAEA experts were doing and waved off both the interpreter and the observer. From then on, the questions and answers were direct expert-to-expert with the interpreter and observer sitting in the corner with nothing to do but sulk.

The week-long set of meetings served to more or less validate Tennessee's original version of the scenario but was useful to help fill in more detail and to develop a time line of the events. For Tennessee, there was still a nagging question regarding the so-called operator error. Setting the stop-power switch was clearly written into the test procedures and was of such importance that it is difficult to believe that such competent and experienced operators would make this error. In risk analysis space, it would be classified as extremely unlikely. Their explanation that it was due to a mix up resulting from the unusual shift change was not exactly a satisfying answer for Tennessee.

Most of the final day was spent with the other IAEA experts to put together the overall TecDoc. After its completion and review by the group, they all agreed that it was the definitive report on Chernobyl and would likely be highly regarded.

As Howard and Tennessee were preparing to leave the IAEA headquarters, the three Russians they had met with on their subject area intercepted them. Dr. Demmen said: "we very much appreciated our interactions with you and hope we may somehow continue our connection in the future. Please accept these gifts of two bottles of our best Russian vodka and join us in a farewell drink." After consuming one bottle with toasts and indigenous jokes accompanying each glass, they shook hands as one last gesture of friendship. Tennessee shook hands with Felicitas last. As he did so, he noted a worried look in her eyes and felt the secretive passage of a carefully folded small note.

Later, when he was alone, he unfolded the note. It read... *"Please come to my room tonight. I am in 301 of the Nordbahn. Come around midnight and please do not let anyone see you."*

Of course, Tennessee was much intrigued by this and yet somewhat worried because of the previous distressed look in her eyes. There is no way he could refuse such a request. At midnight, after carefully casing the hallways, he made his way quietly to room 301 and lightly rapped on the door. Felicitas slowly and carefully opened it. "Quick! Come inside."

Felicitas grabbed Tennessee in a bear hug -- her chin barely reaching him at mid chest. "Oh, thank you for coming." This embrace was the nicest feeling Tennessee had in a long time. He didn't want to let go but, finally, he asked: "what is this all about?"

"You have not been told the complete truth about the accident. Our group was warned not to mention what I am about to tell you. I am risking being labeled a traitor to the USSR, but I cannot in good conscious let this pass without the truth being out. If you are willing to hear it, we could both be in great danger. I sensed you were a man of character who would not deny the truth even at risk of your own life. You may leave now if you choose but I am hoping you will stay and hear me out. Before you decide, there is another request that I have. After disclosing this, I cannot stay in Russia. I am asking you to help me get asylum in the U.S. in exchange for my efforts on their behalf."

"I was hoping this meeting would be for other reasons but, if I can be of service to you, I would be honored. I am willing to risk the danger and I am certain asylum can be had."

Felicitas noted that Tennessee had been right on target with his concern about the operator error. "That is one of the reasons I have come to you with this. The test as outlined in the scenario you developed was pretty much correct. However, the test had two purposes…and the second is secret and was not disclosed.

It was not an operator error that kept the reactor from stopping at the 700MW level. This would have required having the control rods appropriately placed in the core. The reason they were not so placed was because the soviet military has developed a sinister electronic radio operated device that can be operated remotely outside any digitally controlled nuclear power plant and can take over control of the scram system as well as the emergency cooling system. They were first going to check if the remote device could actually be used to withdraw the control rods all the way and then they would re-insert them to get back to the 700MW level. The remote device was also intended to shut off the cooling pumps. With the control rods fully withdrawn and the cooling severely reduced, they believed they could avoid trouble by reinserting the control rods. The control rods did try to insert on demand from the remote device but they were so slow that they were completely ineffective because the conversion of the low amount of cooling water into steam had put the void coefficient in complete control of the power leading, of course, to the super prompt critical accident"

Tennessee gulped "what was the purpose of the secret control device"?

Felicitas noted that the USSR was fearful of Reagan's "Star Wars" program and they realized that it could bankrupt the USSR to try to match it. "Secretly using this device on a U.S nuclear plant to call for a withdrawal of the control rods and a reduction in the cooling flow would result in a TMI like accident."

Tennessee realized that the reactor would go into a power ramp that would melt and disperse fuel into the water causing a steam and/or a Hydrogen explosion.

"Yes" replied Felicitas. "But it would not be as serious because the steam explosion in the U.S. plant would be milder and would not breach the external containment. This would prevent release of radioactivity. However, it would certainly destroy the reactor and cause panic among the surrounding population. The result would be so costly to the U.S. economy that it would be like a depression. This could prove to be a neat inexpensive way to win a "cold war" without there being any direct casualties."

Felicitas further noted "the KGB is already in the U.S. with a second device. They are just waiting to hear the results of this test before they use it. We must stop them before they carry out their plans."

Tennessee said "let's not waste any more time. Get your things together now while I put in a call to the embassy. They will send a car to pick us up right away."

When Tennessee explained to the embassy staff that he had an important Russian scientist that wished to defect to the U.S. with significant security information, they were all too pleased to agree to pick them up at the hotel and arrange travel back to the states. The embassy staffer explained "we will pick you up around back of the hotel so as not to be seen. We will be in a black town-and-country limousine. We will be there in about 15 minutes."

Tennessee and Felicitas waited an appropriate amount of time and crept down the stairway to avoid the elevators. As they exited the back door onto the street, they were greeted by the so-called observer. "Are you two going somewhere?"

Felicitas blurted out "watch out Tennessee. He is with the KGB and will stop at nothing to prevent our escape."

Without hesitation, Tennessee waded into battle happy to have this first opportunity to use his years of karate training. It seems, however, that the Russian also had combat training. As Tennessee attempted a chop shot to the neck, the agile Russian grabbed his arm in mid-flight and deftly slammed him to the ground. His training must have been in jujitsu. Tennessee jumped back up. "To hell with karate" he thought as he regressed to his mountain fighting style. He slammed a right jab into the Russians soft belly which doubled him up. With a left upper cut and a right cross, Tennessee finished him off.
"He will be out for a few minutes. Let's grab our ride."

Chapter 5: Needle in a Haystack

Back in the states, Tennessee arranged with the State Department to have a safe haven for Felicitas. They then quickly met with the CIA and the FBI to explain the issue. These G-Men understood the gravity of the problem, but were somewhat at a loss as to what to do about it. As one of them said, "with one hundred and four reactors out there, searching for this device will be like looking for a needle in a haystack."

Tennessee offered to help. "I have some friends who may have some ideas as to how to proceed."

"By all means, but keep us in the loop as to what you are going to do."

Tennessee arranged a meeting with the F5 group at the State Department building. He quickly explained the problem. Tennessee added "we need a brainstorming session on how best to find and disarm this device before it is used."

Robert Sterling commented "a good way for us to proceed would be to put ourselves into their shoes. If we were the ones that were doing the deed, how would we proceed?"

"Good thought added James Arnold. If I were doing the deed, I would look to find a target reactor that would have both the maximum impact and would also provide a good chance to actually succeed."

Tennessee mused "the maximum impact would be at a site that has the largest surrounding population in a highly visible section of the country."

"Yes" chimed in Howard. "I would think that would narrow the choices down to about five sites. I would suggest these would likely be Salem, Indian Point, Braidwood, Millstone, and Byron.

"Good choices", said Tennessee "and I happen to know that each of these have updated their earlier analog control systems to digital and would therefore be plants vulnerable to such an attack. Since there are five of us and we are focusing on five plants we should each take one."

The drawing out of plant names from a hat gave the assignment of Indian Point to Tennessee and Salem to Howard.

James Arnold asked, "now that we know which plant each of us have, how are we to proceed to find this needle in a haystack?"

Tennessee said "I have thought a little about that. We quietly visit our assigned plant and convince them to disconnect their digital control system and reattach the previously relied on analog system. These have been left in place as backup and are still there. Disconnecting the digital control system will protect the plant. We then place a ring of detectors around the plant containment that can rapidly scan all possible radio frequencies that the device might use. When we detect an unauthorized frequency, we triangulate to locate the position of the device. We, then, will have caught those KGBs.

How are we to identify a KGB anyway, asked James?"

Haven't you seen any James Bond movies? The Russians always wear black hats and dark suits and speak in a funny accent."

"Thanks, sneered James, where do we get these detectors?"

"I happen to know that such detectors already exist. NASA uses just what we need in their Search for Extra Terrestrial Intelligence (SETI) program. In this program, they scan the range of radio frequencies over different segments of the sky. I'll have Vic Parnelli of Oak Ridge see if he can borrow some of their extras or have more made up. Because of their limited pickup angle, for our 5 plants we will need about 25 of these. I figure the plants will require about 2 weeks of shutdown to make the change over back to analog control. During that time, the digital systems will be disconnected and the plants will not be vulnerable to the induced accident."

"Good plan," said Simon Trent. And, in his best managerial capacity, he added "we also have to first get approval from the Nuclear Regulatory Commission as well as get them to issue a regulatory order to these plants requiring them to give us access to their plant and to allow us to implement the plans."

"You are right," replied Tennessee "and we have no time to lose. As we are already in Washington, let's get the Secretary of State to call the Chairman of the Commission and arrange an urgent security meeting with us today. After that is taken care of we, must to get started on the change overs."

The meeting at NRC headquarters in Rockville was not held in the usual Commissioners public conference room. Because of the high security requirement, it was held in the smaller completely secure conference room on the eleventh floor. The F5 group was accompanied by a high-ranking state department employee to give them the needed credibility. The politically and technically astute commissioners listened to the explanation of what this was all about. They then excused the group so they could discuss this among themselves. Although the Chairman has the authority to issue such orders in an emergency, he is generally reluctant to do so without the agreement of at least two of the other four commissioners.

When they recalled the group together, the Chairman had what Tennessee would call a "hang dog" expression on his face. The Chairman spoke. "The Commission appreciates the concerns of public representatives such as you with the safety of the nuclear plants. We have, however, decided that this is not a safety issue but mostly an economic one. By law, for us to issue such an order we would have to make a "backfit" regulatory analysis to show clearly that the risk to the public outweighs the cost. To shut down five of our most highly powered electric plants for the estimated two weeks it would take to re-establish the analog control systems will cost the plants billions of dollars whereas the public is at very little to no risk. Besides, how can we be sure the target is actually one of your selected five plants?"

"There is no way it can pass the backfit test. We are sorry but we must refuse your request for such an order."

Simon Trent chimed in, "we appreciate your legal constraints, but I have an alternative suggestion. Could you arrange for each of us to have a meeting with the owner of his assigned plant to see if we can convince them to voluntarily make this temporary change in an effort to protect their investment?"

With a smile, the Chairman said "of course, we will do this immediately. When do you want to visit each plant?"

"Today will be soon enough," said Tennessee.

"Tomorrow is the best we can do," said the Chairman.

"Tomorrow it is then."

The F5 then headed to their respective home offices to make plans for the next day's visits to their respective assigned plants.

Tennessee arrived at the Indian Point site near New York City early the next day. He was met at the site boundary entrance by John Gaston, the plant vice president of operations. "This must really be a serious security issue for the NRC Commission Chairman himself to call to set up this meeting."

"It is," said Tennessee. "I will explain it all inside if you can get together everyone in authority for operation of the plant for a meeting. Note, this is not an Indian Point performance issue. It is an external threat.

Inside the plant, it took about half an hour to round up all the required parties. Tennessee started the meeting by stressing the need for high security. He was informed that the meeting room itself was a high security one. "I must also get each of you to agree not to discuss this meeting with anyone even the rest of the plant personnel."

After receiving unanimous agreement, Tennessee explained in detail the threat and how it had become known to the U.S.

John Gaston wheezed, "whooee, what are we expected to do?"
Tennessee then outlined the plans for shutting down the plant and switching back to analog control for indefinite operation until they traced the device and caught the Russians.

"Look," said Gaston, "the likelihood of our plant being the target is like one out of a hundred. We can't afford the cost for this changeover under those odds. We believe we can just take our chances on this."

"In that case," replied Tennessee. "Will you at least allow us to place a ring of detectors around your plants containment so that, if you are the unlucky one, we can still catch the culprits so they can't do it again?"

"Of course, we can allow that. You will have to use our technicians, however, to monitor these"

"That's no problem," replied Tennessee. We will get started as soon as we find out the status of the detectors."

With this alarming change in tactics, Tennessee called the other members of the F5 to check on their progress.

The responses from the other plant management teams were similar. None of them are willing to voluntarily shut down. The exception was the Salem plant which, by coincidence, had scheduled a planned shutdown for refueling at this same time frame and believed they could work the changeover into their schedule. Tennessee pressed the group to at least get permission to place the ring of detectors around the containments.

Back home in Oak Ridge, Tennessee was much bothered. "We are now in a condition of catching the Russians only after they will have accomplished their main objective. Such a strike at even one plant is likely to give them the intended economic disaster results."

On reporting the situation back to the CIA, he was informed of an additional tactic that might be useful. The CIA agent, Wayne Alger, had the following news for Tennessee:

"You may not be aware, but we now have the satellite surveillance capability to continuously monitor any and all movements around each of our 104 nuclear reactors. With enough lighting placed around the exclusion zone, we can see an intruder and perhaps intercept him before he can do the damage. We have already started implementing this plan and should have everything in place in about a week."

Tennessee was much heartened by this good news and was in better spirits when he checked in with Vic Parnelli to determine the progress on getting the detectors from NASA.

Vic said, "NASA was all too happy to help us out. They can produce these detectors almost on an assembly line basis. They promised to have the needed 25 units within about10 days. Oh, by the way, the cost will be about $100,000 each for a total of 2.5 million. Do you have that much in your 401K?"

"With the pay scale here at the Lab, I'm lucky to have $401 in my 401K!" He added "this is a national security issue. I'm sure the State Department will finance this."

Vic replied, "well you had better let them know right away and, by the way, get it down in writing."

Tennessee called his new contact at the State Department to discuss the financing issue.

On answering the phone, the State Department contact blurted out, "Tennessee, we are in a panic here. Felicitas has disappeared. She left a note that said she was certain that the Russian KGB had found out where she was staying and she feared for her life. The note said she is going to find a safer hideout on her own. We have no idea where she is."

This news greatly disturbed Tennessee as he had begun to view Felicitas as more than just a fellow scientist. He had never met anyone else who he longed so much to hold. He sensed that Felicitas might feel the same toward him, but under the circumstances, he had not yet acted on his feelings. He had no idea of how to start looking for Felicitas. He would just have to hope that she knew what she was doing and was safe. It now looks like he must find two needles in this haystack.

Chapter 6: Confrontation

Felicitas knew where Tennessee worked and that the Oak Ridge location for the Manhattan Project was chosen because of it being remote and hidden away between the Cumberland and the Smoky Mountains. She speculated that this area might be a better place to hide out from the KGB and, besides, she would then be nearer to Tennessee Crockett which had been much on her mind.

She had escaped Washington in the car that the State Department had left with her in the so called "safe haven" in case she needed emergency transportation. She had found out from friends in the Russian embassy that the KGB knew her whereabouts. She deemed this sufficient to be characterized as an emergency.

It took about eight and a half hours to drive the 500 or so miles down 1-81 and 1-75 from DC to Oak Ridge. She utilized the cars GPS device to lead her to the address of the new log house on Tennessee's 120-acre farm on the outskirts of Oak Ridge.

When she arrived, her first impression was "Looks to me like a good place to hide." Tennessee had, somewhat intuitively, situated his house in such a way that it could be defended against any unknown attack. It sat near the top of the hilliest part of the acreage with its back against a mini-mountain that was completely forested and rose steeply to a height of about 2600 feet. The front had an entrance gate equipped with intrusion detectors and a long driveway leading to the house. The acreage in front was cleared into pasture land providing complete visibility in all directions.

The long drive down from D.C. resulted in Felicitas arriving at a time in the evening when Tennessee should be home from work. So, she took the chance to drive directly up to the house and park the car in back out of sight. She knocked on the door and held her breath while awaiting a response. On opening the door, Tennessee beamed with surprise. "Felicitas, I've been so worried about you."

The immediate embrace was continued without either of them wanting to let go. "Oh Tennessee," Felicitas whispered, "I've wanted this since I first met you. I had to see you again."

They sat and talked until late that night catching up on each other's past lives. Considering the huge differences in their birth places, their early upbringing was still amazingly similar--hers in the Republic of Georgia and his in Tennessee. It turns out to be a small world after all. While sorely tempted, they decided it was too soon to consummate their love. The best things in life are worth waiting for.

"We need to plan how to make sure you are safe here," Tennessee said. "If you can find me as easily as you did, so can the KGB." He added "can you shoot a gun?"

"I am a crack shot with a pistol."

"Good," replied Tennessee. "Being a typical Tennessee hillbilly, I have several pistols, shotguns, rifles, BB guns, and sling shots here in the house."
"You should keep a loaded pistol with you at all times and so will I."

"I think staying here in the house all the time will make you a sitting duck. A moving target is harder to hit. We need to keep moving, but varying our route where ever we go."

Tennessee informed Felicitas of the plans for the detectors and the satellite surveillance.

Felicitas thought these were excellent plans that have a good chance of being successful. "I want to do whatever I can help."

"I will need to go to Indian Point as soon as NASA has five detectors ready. You should accompany me. That way we can watch out for each other."

Three days later NASA called Vic to inform him that they had constructed 10 of the detectors, thoroughly checked them out, and deemed them ready to go.

As there was little time to lose, Tennessee and Felicitas chartered a small jet out of the Knoxville McGhee-Tyson airport and flew to Houston to pick up the detectors. Five were to be used at Indian Point and five at Salem. After picking up 10 units from NASA, they flew to New Jersey and dropped five off at Salem. The next step was to continue on to Indian Point in the jet. The entire trip, including renting a truck, driving to the Indian Point site, and placing the detectors inside the secure area took most of the day.

After dropping off the detectors, they headed for a local French restaurant, "Le Petit Fruite de Mer," that had been recommended to them by one of the Indian Point staffers. In route, they realized that a dark Cadillac sedan was closely following them. As a test, they quickly turned onto a side road, went about 50 yards, and did a U-turn. The sedan did exactly the same thing.

"They must have found out where we are" said Felicitas. "Can we shake them?"

Tennessee gunned the truck and took the first entrance ramp onto the interstate. In the mirror, he saw that the sedan did the same thing. Going about 80mph, Tennessee attempted to weave in and out of the traffic. The underpowered truck was no match for the sedan which still followed closely. At the last possible moment, Tennessee quickly swerved into the first exit ramp hoping the sedan could not react fast enough to make the turn. It was a false hope. The sedan stayed with him. Not being familiar with the area was a handicap. 'If I were doing this in Oak Ridge, I would know exactly which routes to take to lose them." At this point, the sedan was almost touching the rear of their truck.

Tennessee said to Felicitas "get your gun ready. I'm tired of fooling around with those guys. Be sure your seat belt is fastened and brace yourself."

With that accomplished, he slammed on the brakes causing the sedan to "rear end" the truck. After the collision rolled to a stop, Tennessee jumped out of the truck and raced back to the sedan. Air bags had activated and temporally disoriented the driver and the front passenger. There was also a rider in the back seat that apparently had not had his seat belt fastened. He was already unconscious.

Tennessee dragged the driver out of the car and gave him a disabling karate chop across the neck. He then pointed his pistol at the passenger and said, "get out now." Another quick karate chop and this one was also out.

Tennessee and Felicitas quickly removed the shoes and pants from the supposed KGB agents. They then tossed these into the back of the truck.

"The truck doesn't look too worse for the accident," Tennessee observed. "Let's get out of here before the police arrive. I'm still hungry."

The next morning Tennessee picked up a local newspaper on the way to breakfast at the Cracker Barrel. He chuckled as he read on page 6 the paper's account of a strange hit-and-run accident.

"The police reported that a badly damaged Cadillac sedan was apparently involved in a collision with an unknown vehicle that had disappeared. The occupants of the sedan were found unconscious outside the vehicle with their pants and identification missing. They claimed to have been mugged and robbed by a gang of thieves. Later, however, the police located the missing pants and wallets containing identifications at the side of the road some half mile away. They thought it strange that no money or credit cards were missing. As best they could determine, the occupants were illegal aliens from Russia that had come in from Canada in a rented car. The police surmised they may have been the victims of a rival gang of some sort. They are being

held pending arrangements through the Russian embassy to have them deported."

Tennessee said, "I don't think we have to worry about that particular bunch for a while." Felicitas merely smiled. After a hearty breakfast at the Cracker Barrel, they proceeded to the Indian Point Nuclear site. John Gaston was awaiting them, "we have our technicians ready to install the ring of detectors whenever you are ready to show us just where you want them."

"Just evenly place them completely around the containment. Picture the minimum diameter circle you can make with the core as the center and the detectors evenly placed on the circumference with an angle of 72 degrees between each."

"That should be easy enough," said John. "There were instructions packed with the detectors explaining the electrical hook up and use so I think we can take care of the rest."

"Good," said Tennessee. "I think you should have your security force involved in the triangulation and maybe you should conduct a drill on how best to quickly get to the location so determined."

"Will do" said John.

Tennessee continued, "I presume you are in constant contact with the CIA with regards to the satellite surveillance?"

"Yes, as are all of the other 104 reactors. Our security people are up on this aspect too."

Tennessee noted "I still have to collect the other 15 detectors when they are ready for the other plants. I guess there's nothing else for me to do but wait."

Chapter 7: The Capture

After the remaining detectors were put into place at the selected nuclear sites, Tennessee and Felicitas awaited further developments back in Oak Ridge. Three weeks had now elapsed since they left Vienna. Even though there was much uncertainty about whether or not they had done enough to prevent the induced accident and much suspense regarding any additional KGB activities, Tennessee and Felicitas thoroughly enjoyed this free time together.

They spent the time exploring both the nearby Big South Fork Wilderness Area and the Smoky Mountains. Tennessee even taught Felicitas how to fly fish for native brook trout on one of his favorite streams. The brook trout still survive only in the very upper stretches of these Smoky Mountain streams because they have lost the competition to rainbows and browns in the lower stretches. Consequently, Tennessee and Felicitas had to hike at least 3 miles up the stream before even starting to fish. Tennessee was somewhat amazed at the endurance of Felicitas on these arduous up-hill climbs.

Much to Tennessee's dismay, however, Felicitas seemed to enjoy Gatlinburg and Pidgeon Forge more than the fishing and hiking. She even liked their visit to the Dollywood Art Festival. Tennessee thought, "I guess I can't expect her to be perfect. After all, she is a female."

After 5 weeks had passed and having heard nothing from the nuclear plants, they began to wonder if all their efforts to thwart the attacks had been for naught. Perhaps the KGB had somehow learned of the defensive measures taken and decided against proceeding with the action. Or perhaps they were just biding their time.

Exactly 37 days passed before they received a call from Howard Clemons who had Salem as his assigned plant. "Bingo! We got them and without any damage being done to the plant. It may interest you to know that they might have succeeded had Salem not had its digital control system disconnected according to your idea. They were spotted by the satellite surveillance technique but they had activated the device before the security guards could reach them.

We also got the device which, if you don't mind...or even if you do...l want to take back to Sandia with me to determine exactly how it works. We may make one that would be specific for the Iranian production reactor and pass it on to Israel for whatever use they want to put it to."

Tennessee cited, "I guess 'extremism in the pursuit of justice...''

"Does that mean it's OK with you?"

"Of course, but you need to check it out with the State Department and DOE."

"Do you think I'm wet behind the ears? I know all that."

"My mistake. I thought you only knew how to push around theories and non-linear differential equations."

"What do you think bureaucrats are? They fit into that category…especially the non-linear part."

"Well, good luck with that. And don't bother me for a few months...Felicitas and I have plans."

"I know. She is going to teach you Russian and you are going to teach her Tennessee -- both of which are foreign languages."

"Yeah, something like that. I plan to make her an honest person and a U.S. citizen by marrying her and then we are going to take a long honeymoon."

"Are you crazy man? You have been a bachelor for 35 years. Why change now?"

"I guess I just never found anyone who suited me and besides I'm tired of being a virgin. My biological clock is not only ticking, it's chiming the midnight hour."

"Yeah, right! Do you think she bought that cock-and-bull story about you being a virgin?"

"You've got it all wrong. This was her idea not mine. But I am willing to go along with it."

"Well, you have my blessing. I guess the F5 is now going to be the F4…or will it now be called 4F?"

"I'm just getting wed...not dying."

"Same thing. But we shall see how much you still participate."

"I still have to make a living unless you want to donate some substantial money on a regular basis."

"I couldn't give you much. Sandia doesn't pay any better than Oak Ridge Lab."

As neither Tennessee nor Felicitas had much family nearby, the wedding took place in the little commercial chapel in Gatlinburg in the midst of December with only the presence of the Chaplin, a hired witness, and another couple awaiting their turn. In Tennessee's luggage were two round trip tickets to the Bahamas.

Tennessee was lying leisurely on the tropical beach admiring his new bride in her little bikini. His only coherent thought was, "it's true, good things do come in small packages."

A page, trying with limited success to not stare at Felicitas, approached Tennessee. "Dr. Crockett, you have a phone call up at the pagoda."

Tennessee sighed, "I wonder how they tracked me down? Not even Katie knew where I was going."

Tennessee told Felicitas about the call and then ambled back to the pagoda. At the indication of the telephone by the bartender, Tennessee picked up the receiver and said, "whoever you are, Tennessee said to tell you he was not available and had left for parts unknown."

"You can't fool me. I would know that Tennessee twang anywhere," said Howard on the other end of the line.

"How did you find me? I didn't leave any forwarding address with anyone?"

"It was easy," replied Howard. "I just had NASA use their global positioning system to focus in on anything that still had wormwood debris on their skin and was being bedeviled by evil spirits."

"I should never have told you about that," answered Tennessee. The only reason I did was because of the eerie connection between Revelation's wormwood and Chernobyl. What is it you want that is so important for you to hassle me on my honeymoon?"

"Nothing. I just wanted to see if Felicitas was tired of you yet and wanted to give me a chance to take your place."

"You know I'm a black belt in karate, don't you? I still owe you for that thing you did to me in that Vienna restaurant -- whatever it was."

Howard got a little more serious. "I just wanted to let you know that we got another one. They had another of those devices and tried it at Braidwood. Fortunately, this time the surveillance system was sufficient. We caught them before they could activate the device."

"That is a surprise," said Tennessee. "I hope there are no others to worry about."

"I doubt there are any more," replied Howard. "But, I also want to know when you are returning to Oak Ridge. The State Department has another chore for you and me."

"We have about another week here. What is this new chore?"

"You remember I said we might be able to make one of those Russian Chernobyl devices that would be specifically designed to operate on the Iranian production reactor when it is completed. Well, we think we have succeeded in this. The State Department, in their infinite wisdom, doesn't want Israel to have the say-so on the use of this. Consequently, they have "suggested" that you and I secretly work with Israel to make appropriate use of the devise."

"Why us" replied Tennessee?

"Because, they think we are the only ones that know how to use it and that we are also stupid enough to accept such an assignment."

"Sounds like fun," said Tennessee. "I'll see you in about a week."

The last night of their honeymoon Felicitas wanted to take a romantic moonlight walk on the beach. There was a warm breeze blowing with a full moon above. As they walked along the now deserted beach, Felicitas noted that Tennessee seemed to be admiring her gorgeous five feet four figure. "Do you like what you see?" she asked teasingly.

"What I can see of it," replied Tennessee in somewhat of a cryptic manner.

A bit piqued and on a sudden impulse, Felicitas removed her shorts and halter. She twirled around completely nude in front of Tennessee. "Is that any better," she continued to tease.

All that Tennessee could say, after he got his eyes refocused, was "I believe my life is going to be a lot more interesting than it has been up to now."

The subsequent lovemaking on the beach and the leisurely stroll back to their rented cottage was a fitting end to their honeymoon.

Back in Oak Ridge, Tennessee had to reluctantly leave Felicitas at their log home while he went to Albuquerque to make plans with Howard to carry out their task of applying the Chernobyl device on an Iran production reactor. As the device used large car-like batteries for its power source, it was quite bulky and heavy. It was almost all Tennessee and Howard together could lift.

Their plan was to load the device into Howard's pickup truck, and take it to the Albuquerque International Airport. There they would turn it over to the Israel team who was to have flown in via a private jet just for the purpose of picking up the device and to transport it, along with Howard and Tennessee, back to Israel. After being trained in its use, the Israelites would then fly Howard and Tennessee back to the states.

On leaving the Sandia reservation with their package they noticed that, once again, there appeared to be following them a dark sedan with strange looking occupants. As best he could tell by observing in the rear-view mirror, Howard noted that "they have swarthy skin, dark hair, beards, are wearing turbans, and have an evil look in their eyes."

Tennessee said, "you would think by now that Arabs could do a better job of disguising themselves."

"I don't know how in hell they know what we are doing," said Howard.

Tennessee replied "well we know the Russians are in cahoots with the Iranians. They have been supplying the components and fuel for the plutonium production reactor that the Iranians are building. The Russians must still be keeping tabs on us and have informed the Iranians." Howard remarked "I remember how you escaped from the Russians at Indian Point. I do not have any desire to induce a wreck with my brand-new pickup truck. I suspect that raunchy looking bunch behind us hasn't any insurance and I can't afford a large repair bill."

Tennessee snorted "that's what you get for having a brand-new truck. You should have a 15-year-old Dodge like mine. If so, you wouldn't care if it got a few more dents and, besides, they are built like a tank."

"To each his own," replied Howard. "Just what do you suggest we do as an alternative to banging up my truck?"

"They probably have machine guns and bazookas and all I have is this little 22 caliber pistol. We are certainly outgunned."

"Thanks for those encouraging words" said Howard. "Haven't you got some better plan than that?"

"When I play basketball, I have learned that a good offence is often better than a good defense," mused Tennessee. "Why don't we surprise them?"

"How?" asked Howard, a little intrigued by this thought.

"Make a quick U-turn and head straight back toward them. Let's see how good they are at chicken."

"Are you crazy? Those are the idiots that strap bombs on and blow themselves to smithereens just to receive 72 fictitious virgins in nirvana."

Tennessee said, "I have often wondered what terrible things those virgins did to deserve such a fate. And after they are no longer virgins, then what?"

"This is no time to be philosophical," said Howard.
"I think they will chicken out," said Tennessee, because they want to be sure to get their hands on this device. They can't do that if they are in nirvana fooling around with a bunch of virgins."

"O.K." said Howard. "What do we do after they chicken out?"

"Leave that up to me "After they chicken out, just stop the truck and let me out. If I take too long to get back you keep on going to the rendezvous with the Israel team and go ahead with them to Israel.

"Are you sure" asked Howard?

"Don't worry. I can take care of myself" said Tennessee.

The quick U-turn by Howard took the Arabs by surprise and it took them a little distance of travel before they could do the same. As they made their U-turn, Howard made another U-turn and headed straight head-on towards them.

There was a look of panic in the eyes of the sedan driver. He was not a good chicken player. He quickly veered instinctively to his right careening off the road and into a waiting tree that providence had placed there just for that purpose.

On seeing this result, Howard said "I don't think there is any need to let you out, Tennessee."

"You are right" replied Tennessee. "They are not going anywhere now. Let's go on to the airport."

Howard added, "I don't know if you are the smartest man in the world or the dumbest, but that was a damn good idea on your part back there." "Aw shucks" Tennessee drawled. "I didn't know you cared."

Chapter 8: Turning the Table

Back in Oak Ridge, Tennessee was glad that he did not have to accompany Howard to Israel. He was anxious to spend more quality time with Felicitas.

At the Oak Ridge Lab, Tennessee went about setting the stage to have Felicitas become either a visiting scientist or a permanent member of the Laboratory. There was no sense in wasting a fine scientific mind like hers on just homemaking and she was anxious to get back into serious research. In addition, they could use the extra money.

Felicitas felt very much at home in East Tennessee. The locale reminded her of where she was raised in the Caucasus mountain area in what was known as the Republic of Georgia. The culture in that area did not seem that much different from that of East Tennessee. She found the Tennessee people to be warm, friendly, and accepting. That is, of course, with the exception of the ones that insisted on shooting each other over some real or imagined infidelity. But you find those types everywhere except maybe in Japan or Switzerland.

No wonder, she thought, that Tennessee insisted on continuing to live here even though he had many potential lucrative offers elsewhere. She thought "I would be happy to live here the rest of my life."

After some time had passed without any more attacks on a nuclear plant or on Felicitas' life, Tennessee and Felicitas were beginning to relax and enjoy their new jobs and each other. Felicitas, however, was concerned. "My parents still live in Georgia and I haven't heard from them in months. They do not own a computer or telephone and I have written them more than a dozen letters without any reply. On asking Josh Hammer of the State Department about this, he filled me in that the Soviet Union has been suffering from inflation, supply shortages, and severe social and economic problems that have been exacerbated by the Chernobyl event. As a result, there have been pro-independence movements in a number of the soviet republics-Georgia among them. Georgia is one of 5 republics that have declared independence from the Soviet Union and have actually held elections to form independent governments.
The Russians have sent troops with tanks and apparently have cut off all communication with these republics in an attempt to stop these independence movements. The Russians can get brutal under these circumstances. That is why I am so concerned with the wellbeing of my parents. Tennessee, is there anything we can do?"

"Perhaps," replied Tennessee. "The NRC and DOE have developed a Memorandum of Understanding by which they are putting together a team of experts to see if there is anything that can be done to improve the safety of the RMBK-IOOO reactors in the short term. I have proposed that the most important short-term fix is to reduce the positive void coefficient to as low a level as possible.

I have suggested that the way to do this is to install many more absorber rods, increase the worth and speed of insertion of the scram rods, and increase the fuel enrichment."

"The first thing for this team of experts to do is to visit some of the Russian plants for a first-hand inspection and to meet with the Russians to discuss possible safety improvements."

"Based on my input so far, I could request to be a member of this group which is to leave very soon on these inspection tours. I could also volunteer you to come along as an expert on the RBMKs and as an interpreter if needed. After our inspection and meeting, it would be an easy trip to Georgia and to your parents' home."

Felicitas thought this was a wonderful idea. "I would like to bring them back to the U.S. with us to live here. Do you think this is possible?"

Tennessee replied "the Russians are not allowing people to leave the country at this time. But, maybe we can pull the ole' switcheroo on them."

"What do you mean" asked Felicitas?

"Let's get some help from the CIA on this. What we need are two agents, a man and a woman that are about your parents' age. We take them along as part of the team-- perhaps as interpreters or in some other support capacity. They go into Georgia with us, trade their identification documents with your parents who then leave with us in their place. The agents can leave later with no one the wiser. We should take round trip plane tickets with us for your parents and just boldly board the outbound plane."

"It sounds very chancy, but it's worth a try," Felicitas tentatively replied.

After about three weeks, all the arrangements had been successfully made with NRC, Oak Ridge, and the CIA and Tennessee and Felicitas were on their way to Russia with the team of experts and the CIA agents.

Felicitas was noticeably nervous. Tennessee reassured her, "believe me, this will work. I checked it out with Howard and he approved. That is good endorsement from the biggest scam artist I know."
 This wasn't much comfort for Felicitas.
Tennessee's suggestions to the Russians on how to reduce the positive void coefficient were well received and assurances were given that they would begin to implement

those changes right away. As the team went off to another plant for a second inspection, Tennessee, Felicitas, and the two CIA agents excused themselves to make what was described as a personal visit and a tour of Georgia. The CIA agents went in a separate car with the excuse of the small size of the rental car and the amount of luggage for the four of them.

Being familiar with the territory, Felicitas knew of the least traveled route into Georgia where their entrance would likely not be overly questioned by the Russian soldiers. At the boundary, they were detained somewhat, with I.D. checks and numerous questions about the reasons for their visit. Felicitas' knowledge of the republic and the proof that she was born and raised there were sufficient to allay the suspicions of the guards and they were eventually allowed to enter.

Felicitas was noticeably excited about being in Georgia again. It had been a number of years since she had left for graduate school and for work at the Kurchatov Institute. She delighted in pointing out landmarks and noting how similar the countryside was to East Tennessee. She looked north eastward toward the Caucasus Mountains and said, "see those. If you didn't know better, you would think they were the Smoky Mountains." Her excitement rose substantially as they approached the old farm homestead where she was raised. Other than needing some tender loving care, it looked very much the same to her. Like many of the farm homes in this area the roof was thatched and the walls were of local stone. It may have been there for hundreds of years and handed down from generation to generation.

As they awaited the response to their knock on the door, Felicitas had to grab onto Tennessee's arm for support. Shortly, when the door was opened, there stood a short, burly looking man barely an inch taller than Felicitas. Felicitas threw her arms around him and said some emotional laden words in the Georgian language which is quite different from Russian. The only word Tennessee recognized was "papa".

Close behind the man was a very nice-looking lady that Tennessee judged to be about 55 years of age. Tennessee could see where Felicitas got her good looks. Go back a few years and she would have looked very much like Felicitas.

After the round of introductions and the great surprise of Felicitas' parents that Felicitas was married to an American, they all sat down in what could be called the living room. It was laborious and slow as the Rochivs did all their talking in somewhat broken English. Tennessee appreciated the attempt. Occasionally, when addressing only Felicitas, they relapsed into Georgian. Felicitas apologetically translated for Tennessee and the two agents.

One of the first items of interest to the Rochivs was "how long will you be able to stay?"

Felicitas noted that they could only stay a couple of days because they would have to rejoin the inspection team. It was at this point that she told them of the plan to have them come with them back to the U.S. to live there.

Mr. Rochiv said "we thank both you and Mr. Tennessee very much for this thoughtful gesture, but we must decline. You see, there are great things happening in Georgia at this time. We were never Russians. We were forced into the Soviet Union against our will. We have declared our independence and have declared ourselves to be a democratic republic." She directed to Tennessee, "you Americans, with your Declaration of Independence and Revolutionary War, will fully understand that we want the same freedom and independence. No, we can't leave now. We are Georgians and we will stay to give whatever support we can to this movement. It would be cowardly for us to leave now."

Although taken aback somewhat, Felicitas fully understood and expressed how proud she was of her parents. "I only wish I could help," she said. "Please hang onto these plane tickets to the U.S. in case you ever need them. They are exchangeable for any flight within the next year."

"You must be hungry," said Mrs. Rochiv. "I'll fix you a real Georgian dinner to celebrate your marriage and maybe encourage some grandchildren."

Felicitas and her mother made their way into the kitchen, while the others stayed in the living room and attempted to make small talk. Although a little awkward, they managed. Tennessee was beginning to like his in-laws very much.

After what seemed like a couple of hours, they were summoned into the kitchen. There on a long ancient harvest table with cane-back chairs, was what reminded Tennessee of the wonderful meals his folks always fixed to celebrate Thanksgiving. The types of food on this table, however, were considerably different from turkey, dressing, ham, and mashed potatoes. They feasted on roast pork, borsch, Russian black bread, whole boiled potatoes, caviar, and a few things that Tennessee had no idea of what they were but they seemed to include cabbage and pickled herring. They finished the meal off with very dark coffee and a delicious cake of a variety unknown to Tennessee.

After helping clean off the table and washing and putting away the dishes, Tennessee and Felicitas were shown around the farm by her father. He noted that they had to sell off the cattle they once raised but they still kept chickens for eggs, a cow for milk and butter, and hogs for the many things that can be made from them that includes meat, sausage, lard, pork rinds, and even soap. They raised vegetables and a lot of potatoes which they sold to the local distillery for making vodka. "Sounds a lot like East Tennessee," commented Tennessee "except we use corn to make our whiskey."

Tennessee and Felicitas retired early that night to her old bed room that contained a straw mattress and the thickest down bed spread that Tennessee had ever seen. As they kissed goodnight, as they always did, Felicitas said "are you sick--you feel like you have a temperature?"

"yes, I do feel like I am coming down with something. I hope it's nothing more than a cold."

"Your temperature is too high for just a cold. Stay here while I get some medicine."

Shortly, she returned along with her mother. Mrs. Rochiv had a bowl full of hot water into which she stirred what looked to Tennessee like dried leaves.

"Drink this," she said.

"What is it," Tennessee timidly asked.

"It is a medicinal herb we pick in the Caucasus Mountains. It will cure anything that ails you."

Tennessee reluctantly downed the foul-tasting brew and laid down to sleep. The next morning, he was amazed at how well he felt and the fact that he no longer had any temperature. He asked Mrs. Roschiv, "what was that wonderful stuff you gave me last night?"

"I believe in your language it is called wormwood!"

finish

www.ingramcontent.com/pod-product-compliance
Lightning Source LLC
Chambersburg PA
CBHW081123240526
45470CB00019B/2932